Cold Storage for Fruits & Vegetables

Storey Publishing

Cold Storage for Fruits & Vegetables

CONTENTS

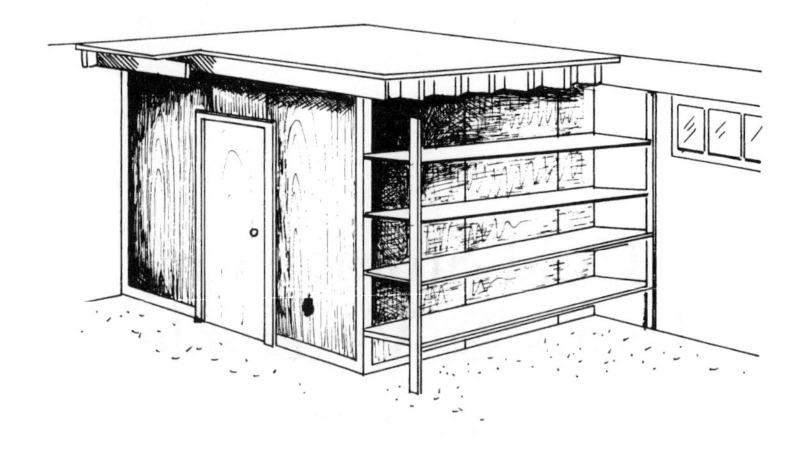

Introduction

Do you want to save what could be wasted garden surpluses? Want to lower food costs, and gain deep personal satisfaction?

Try storing food the natural way, in storage rooms and root cellars, in pits and trenches, or even in the garden.

Here's a way to keep large quantities of food that is cheaper, less work, and far more energy-efficient than canning or freezing. And for many fruits and vegetables, it's the best way to keep them as close as possible to their just-picked or just-dug peak of quality.

It's not difficult, either, if you follow a few simple rules.

One of those rules is that crops are particular about where you store them. They want just the right humidity and temperature. You probably have ideal conditions for some of them in your home right now, and, with little effort, can provide satisfactory conditions for the others.

Onions like **cool, dry** conditions. If you have an attic or an unheated room, you have a good place for onions. The closer to the freezing point the temperature goes, the longer those onions will keep.

Warm and dry conditions are favorable for squash, pumpkins, and sweet potatoes. Many of today's homes have cellars that are fine. The furnace keeps the temperature in the 50–55°F range even during the coldest nights of winter. And the humidity is low. A few wooden racks to keep the produce off the concrete floor and you have good storage space for those crops.

The **cold and damp** cellars of old farmhouses were ideal for storing the root crops, the members of the cabbage family, and potatoes. They were unheated, so in winter were close to freezing, and even in summer remained cool. They were damp, summer and winter, with the packed earth floor giving off moisture, and more creeping in through the stone walls of the foundation.

Reproducing those conditions is possible, and will be explained under storage room construction.

Another rule to remember is that the maximum storage time for vegetables and fruits varies enormously. Ripe tomatoes and peppers can be held for only a few weeks, winter squash will last through the winter, dried peas and beans can be stored for several years. None should be stored and forgotten. They must be checked often, once a week is not too frequent, to make certain conditions are as ideal as possible, and to remove any produce showing decay.

The experienced root-cellar owner senses problem periods. One example: A warm fall will threaten those stored crops by raising the temperature of the root cellar. But if cool air is allowed to flow in during the night, then the room is closed off to hold in this cooler air, the crops may be protected.

A third general rule is that you should plan your gardening so that your storage crops reach maturity — no more and no less — just when you wish to store them, which in most cases is before the truly hard frosts that signal the approach of winter. Most of the crops require only drying in the sun for several days before being stored. They should not be washed before being stored, although dried dirt can be brushed off them, and they should be handled carefully at all times, so bruises do not invite speedy decay.

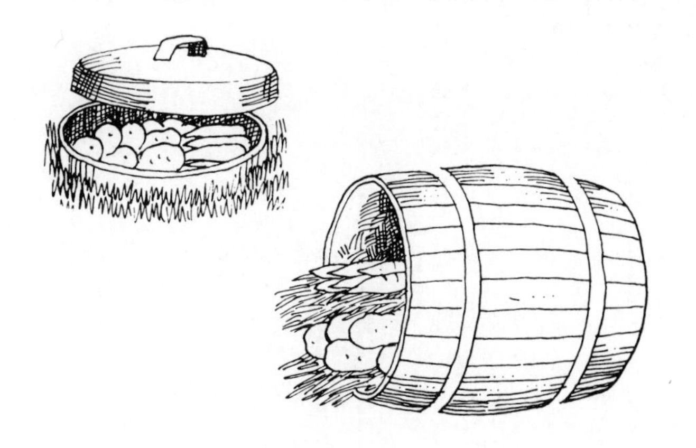

Storage Tips

Here are some tips on storage, aimed at enabling you to store more food for longer periods with less loss.

1. Select the best of the crop for storage. Vegetables should be mature, but not too old, and not damaged by disease or handling. One bad apple will spoil the whole barrel, and the rule applies to most vegetables as well.
2. Don't handle crops for storage when they are wet.
3. Proper ventilation is important. Air should circulate around the stored produce as much as possible.
4. Get your crops in before a hard freeze. Some may not be damaged by heavy frosts; many of them will be.
5. Turnips, rutabagas and cabbages give off strong odors. This is a good reason for storing them in an outdoor pit or mound.
6. Sort potatoes carefully for storage. Use as quickly as possible those that are not mature or whose skins have been broken while being dug up.
7. Most vegetables need to be cured before being stored. Check directions for individual vegetables. Proper curing will do much to increase the maximum storage time.
8. Root crops must retain their moisture to keep their freshness. Store them in boxes of damp sand or sawdust, and keep the storage material damp. Or try perforated polyethylene bags.
9. Most crops store best if kept in the dark. Light is not needed for any of them.

Building Your Storage Room

For most vegetables that need damp, cool storage conditions, a cold storage room or root cellar is the most satisfactory answer.

The homeowner has control over it, control over its temperature and humidity, and control enough to exclude hungry mice that may want to share the feast. The produce is available when he wants it. Carrots can be stored just fine in the garden or in a pit, but retrieving them in the midst of a snowstorm can be trying.

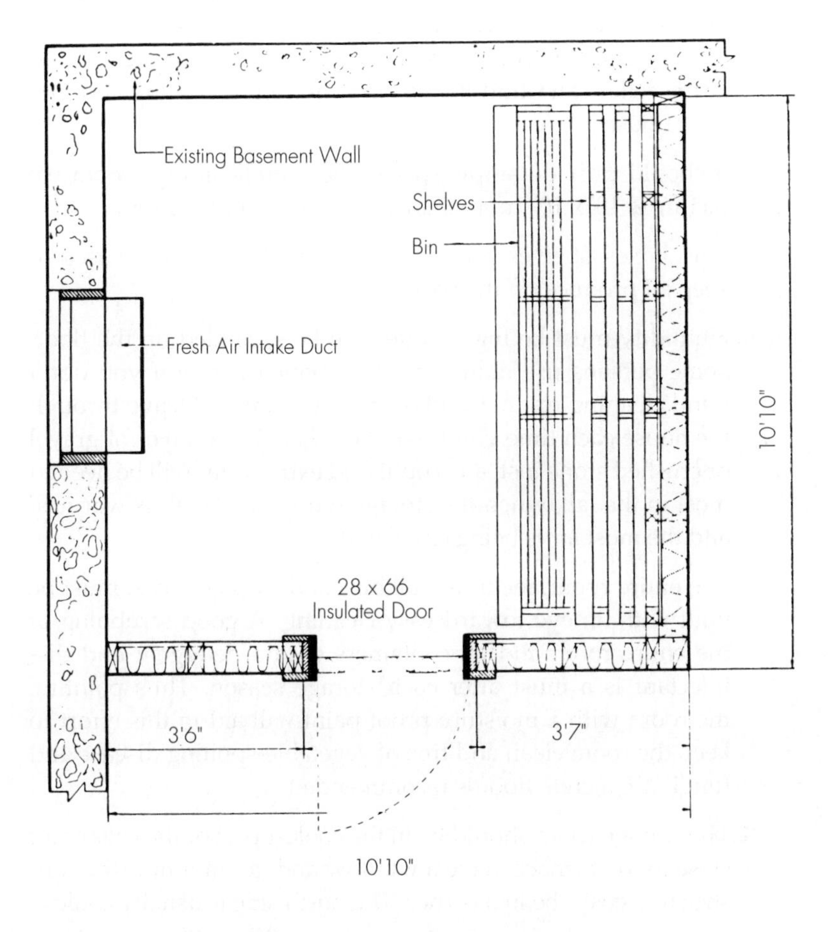

Top view of plan for storage room in corner of basement.

The homeowner must aim for certain conditions in his root cellar:

1. Light should be excluded, but light for use when working in the cellar should be provided.

2. The room should have access to fresh air. That's why we recommend building it to include a cellar window. This air is needed for maintaining low temperatures and to provide the ventilation most crops require.

3. The room should be insulated off from the rest of the house. This is essential if low temperatures are to be maintained. And it's far better for your comfort and heating bills if the cold air of the cellar can't find its way up through the ceiling and into your house.

4. It should provide ample space. The sample storage room we picture is 10' x 10', more than ample for a family of four.

5. Bins, boxes, shelves, and slatted duckwalks should be used to keep all produce off the concrete floor.

6. Humidity must be high. Water can be sprinkled on the floor. Some persons use damp sawdust, which is fine if you don't mind cleaning up that trail of sawdust you will leave through the house each time you leave the cellar. Try a layer of gravel or crushed stone instead. You'll find extra water will be needed most in the fall, when the storage room may be at its warmest and the most air is being circulated.

7. The entire room, including all bins and storage areas, must be built with an eye toward easy cleaning. A good scrubbing of the entire room and all containers with a detergent and disinfectant is a must after each storage season. Thus painting the room with a moisture-proof paint will aid in this effort to keep the room clean and free of vegetable-spoiling disease and fungi. A concrete floor is recommended.

8. The storage room should be in the coolest part of the cellar, not close to the furnace, have a window and, to minimize the construction costs, be in a corner. The north side is usually coolest.

The design we use as an illustration is from the Agriculture Canada Publication 1478, *Home Storage Room for Fruits and Vegetables.*

The first step in construction is to mark off the two walls to be built, then build forms and pour concrete for a 4" x 4" concrete footing. Purpose of this is to raise the wooden wall and its insulation above the level of any water used within the room to maintain high humidity and thus protect them from rotting. Some may skip this step, planning to use care so that the water does not come in contact with the walls.

Use 2" x 4" studs to frame the walls, spacing them properly for the insulation that will be fitted between them. Batts of insulation are cut to fit between studs set on 16-inch or 24-inch centers. If batt type insulation with a vapor barrier is used, that vapor barrier should be on the warm side of the wall. If there is no vapor barrier, use 4 mil polyethylene on the warm side of the insulation. Plywood hardboard or lumber can be used for sheathing on both sides of the insulation.

Use 2" x 2" material for framing an insulated door. Weatherstrip it if necessary to get a tight fit.

Any hot water pipes that pass through the room should be insulated. Circular insulation for just this purpose is available at building supply stores. Hot air ducts, too, should be insulated.

The ceiling should be insulated, too, with the vapor barrier above the insulation and thus on the warm side. There should be no break between walls and ceiling in either the vapor barrier or the insulation. The outer concrete walls should not be insulated.

The ventilation system can be of varying degrees of sophistication and effectiveness. Most simple is a window that is opened when fresh or colder air is needed. This lets in unwanted light as well as air, and does not assure a good circulation of air.

Far better is a system of vents so that air is exhausted from the top of the room, and flows in at near-floor level. Adding an electric fan to the exhaust will increase its efficiency. This system should include dampers so that air circulation can be halted when it is not wanted.

More expensive but more effective is the system shown in the cross-section view of the sample room. This system is automatic, with a fan and louvers set in the upper half of the basement window. These are controlled by a differential thermostat, which measures outside and inside temperatures, and kicks the system into operation when inside temperatures are too high, and outside air is cooler. The system forces the warm air out of the room, and thus air is drawn back into the room through the intake system, with

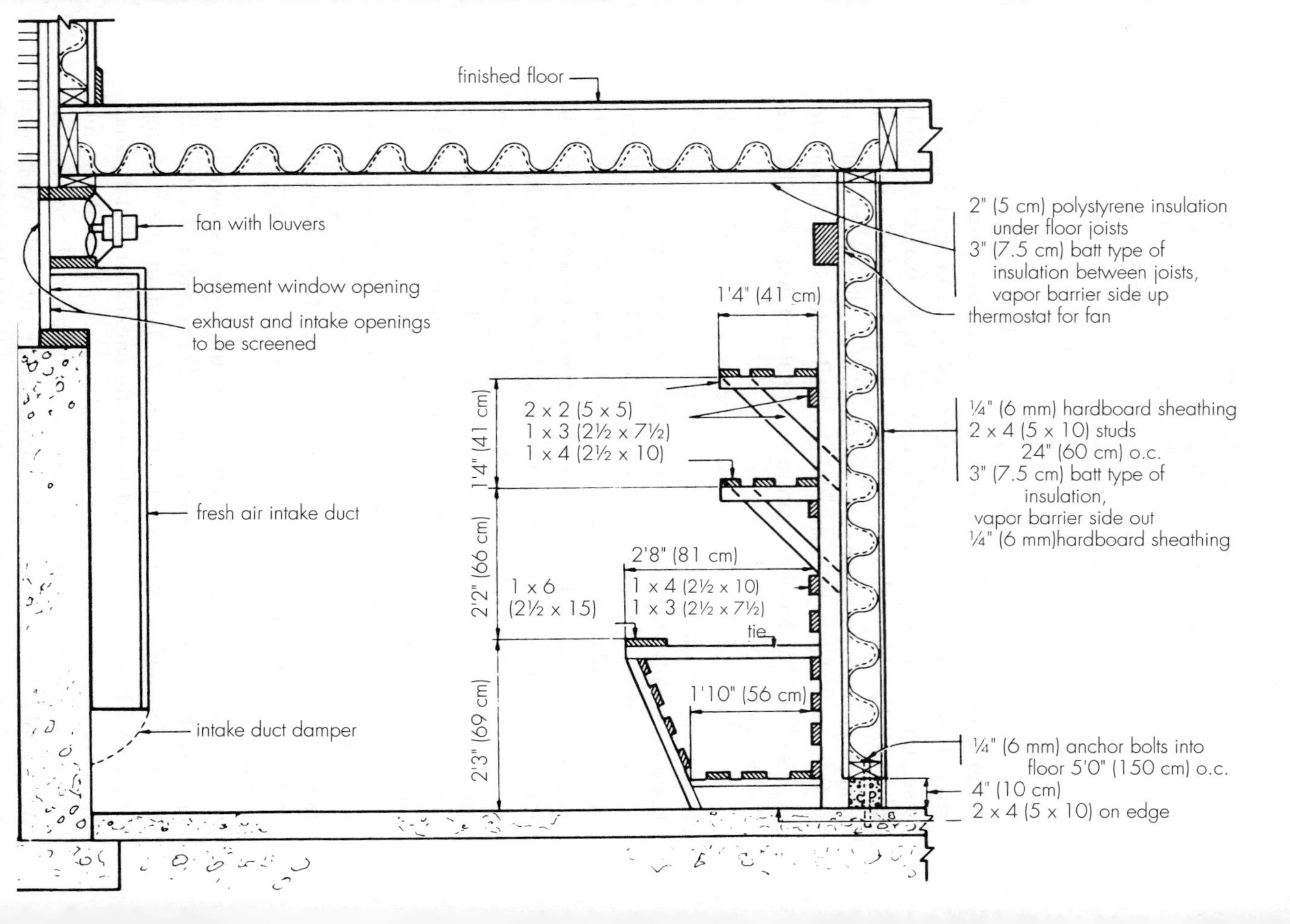

Cross-section view of a storage room.

its outlet extending nearly to floor level. Both inlet and outlet have screening at the window openings to keep out bugs and rodents.

The builder should look ahead to decide what and how he will be storing before he designs and builds shelves and bins. A cardinal rule: Keep all produce, including those stored in boxes, at least four inches off the floor.

There are several advantages to using crates and boxes placed on shelves rather than building bins for crop storage. One, the boxes are cheaper. Two, they're easier to move out and clean in the spring. Three, and probably most important, boxes can be loaded in the garden, then carried into the cellar for storage. This eliminates one transfer of food from one container to another, and thus eliminates one possibility of damaging the produce and thus reducing its storage life. Boxes can be stored one on top of another. Usually it's wise to put small slats of wood between them, to permit better ventilation.

Two instruments are valuable in the storage room. One is the thermometer, to make certain temperatures are as close to freezing as possible, but not below that level. And the second is a sling psychrometer, which measures humidity. It can be purchased from scientific instrument companies.

If too-cool temperatures are a problem in your storage room you may wish to install a low-temperature warning system, which consists of a thermostat in the storage area, connected to a buzzer outside and set so that the buzzer sounds if the temperatures reach dangerously low levels. This equipment and the differential thermostat we mentioned earlier are available at electrical supply houses.

In some storage rooms, where temperature fluctuations are a problem, it may be necessary to insulate the window, at least for part of the winter.

If you construct a storage room, you'll find it is valuable for many uses. It can be used for most fruits, or for vegetables (although you may get in trouble with unwanted flavor transfers if you mix the two). You'll find its weather is ideal for root crops and for potatoes. It's also an ideal spot for short-term storage of other crops. Here's the place for those end-of-season surpluses you've rescued from the frosts. Eggplant can be saved in here for a few weeks, and so can grapes, cauliflower, kohlrabi, cantaloupes, and watermelons. As with the other crops you've stored here, you'll want to keep a close eye on these short-timers, since they should be eaten as quickly as possible, and they will not last long, no matter how ideal the conditions.

Keep Out Canned Goods

If you preserve food in glass jars, perhaps you're planning on shelves in your storage room just for those. That way you keep all of your stored food together. We advise against it. Sure, the temperature is just right — cool — for those canned goods. And they should be in the dark, too, just the same as your stored vegetables. But remember that high humidity. It's fine for vegetables, but it will rust the metal on those cans, and eventually may cause leaks, leading to spoilage.

Instead, while you're building those walls for the storage room, add shelves on the outside for all of those cans and jars of food. Do this, of course, only if yours is a dry cellar.

Pits and Trenches

The storage room is crammed full, and you still have more carrots and beets and cabbages to harvest. Or you haven't time or money or ambition enough to build a root cellar. Or you just want to try different storage methods.

Trenches and pits may be for you.

Be alert for possible troubles, though, since pits and trenches seem to operate on the theory that if something could happen, it will. The spot you choose may look like the Sahara in September, but turn into a dowser's dream of plenty in November. Mice and other tiny animals will find their way in vast numbers into your trench or pit, given so much as a nose-width invitation. And the weather can offer combinations that produce a layer of frozen moisture that's hard as a banker's heart, and will bar you from those carrots or cabbages as effectively as that banker's vault doors.

You've been warned. Now let's try building some of these.

First, you're trying to create conditions in these that are very similar to those in a root cellar. Cool temperatures, but not below freezing. High humidity, but not cold running water. Ventilation. And protection from insects, rats, mice, squirrels, the neighbor's dog, woodchucks, and other creatures with four legs and curiosity or an appetite for vegetables or fruits. Remember all of these, as you try the following suggestions, or your own variations on them.

First, the Pits

They can be used for the root crops (beets, carrots, celeriac, parsnips, the large winter radishes, rutabagas, salsify, and turnips), as well as potatoes. Brussels sprouts, cabbages, kohlrabi, and apples and quinces.

Consider your area temperatures before deciding to try pits. They aren't practical if winter temperatures in your area don't average 30°F or less. If you live in areas of hard winter weather such as New Hampshire, the Dakotas, and the other states in that belt, they can't be used for winter-long storage, since the freezing cold will get to those crops, no matter how high the straw, hay or soil is piled.

Pits are best if they're small. In this way, food enough for a short time can be packed away in one, and all removed at the same time. Try a week's supply in a pit, and remove all of the food at one time. Repacking a pit can be difficult when the ground is frozen, there's a foot of snow on the ground, and the hay or straw has been scattered.

Try digging a hole one foot deep and three feet square in an area of good drainage. Line it with two to four inches of straw, salt hay, hay, or leaves. Make a pyramid of the week's supply of vegetables — carrots, beets, parsnips and others. This pile can be as much as two or three feet high. Don't throw the vegetables in. That may damage them. Place them in position. Now cover them with a deep layer of straw or hay or leaves. A six-inch layer is adequate — a foot-thick layer is better. Add a layer of soil, and the deeper it is the longer those vegetables can be left in cold weather without freezing. If mice are a problem, cover pile with hardware cloth before you add those layers of straw and soil.

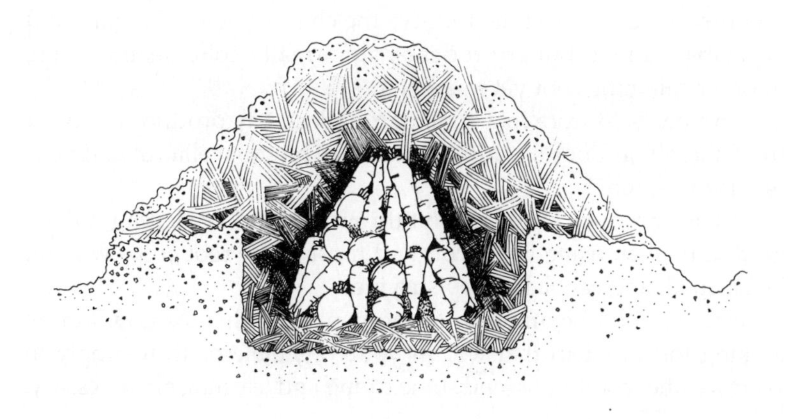

What have we forgotten? Ventilation. Simply letting the straw or hay stick up through the layer of soil will provide enough ventilation for the vegetables. Another method used is to cut both ends off a tin can, and set that in place on the vegetable pyramid, sticking up through those layers above the pyramid. A screen of some sort over the top of the can will prevent it from becoming Mouse Alley.

The final step is to dig a ditch around the pile, and provide a runoff from your ditch so any rain that accumulates can flow away. If you were a Girl Scout or a Boy Scout, you remember how you did this when raising a tent. Same method, and for the same purpose, to keep dry what the ditch surrounds.

Cabbages can be stored in a similar fashion, but with differences that let you get the cabbages when you need them. Cover a storage area with straw or hay, marking out a space that will be wide enough for four cabbages abreast and not touching, and long enough for as many cabbages as you plan to store. Pull up the cabbages, roots and all, and put them head-down on the blanket of straw or hay. Support them by packing hay or leaves around them, so that this insulating material is as much as a foot deep over the roots. Shovel soil over this pile until the soil is at least six inches deep. Dig a runoff ditch around the pile. A few cabbages at a time can be taken from this pile without ruining the layers of insulation protecting the others.

There are many variations on these pits. These include the use of boxes, barrels, and even garbage can sin the pit. There are advantages to using these containers. They tend to protect the produce from mice and water.

There are several rules you should follow if you try the pit method.

One is to use the stored produce as quickly as possible. Just cutting down the storage time reduces the chances of loss of fruits and vegetables due to hungry rodents, abnormally cold weather, rain, or other calamities not yet imagined.

Another is to store in a cool, damp place any produce removed from the pit and not used immediately. The root cellar of course is fine for this, too.

If your produce freezes, don't throw it away. Try letting it thaw gradually, then using it. Carrots and most of the root crops will not be harmed, and cabbages may survive.

Finally, don't be afraid to experiment with this method. You're looking for a system that will fit your vegetable or fruit supply in your weather conditions, and some trying and learning are necessary.

Trenches

Trenches are used for leafy vegetables, including cabbages, celery, and Chinese cabbage This method works very well, although here in northern Vermont, we can't recommend it for all-winter storage. But for a few months — fine.

Dig a trench one foot wide by two feet deep by whatever length you find you need for these crops. If your soil is loose, you may want to prop up the sides of this trench with boards, to prevent cave-ins. Dig up your plants, roots and all. Transplant them into the trench, packed closely together, with soil covering the roots, and the plants deep enough so that their tops are below ground level. Water

them as you replant them, keeping water off the tops. Leave them for several hours at least to settle. Place boards over the top of them, followed by hay, straw, or leaves up to a foot in depth, then cover the entire pile with a piece of plastic.

Visualize what's down there: Your plants have their roots in the soil, so, while they're not growing and prospering, they are getting moisture from the soil. They're protected from the cold, but they do have air. And, when eating time comes, one end of the trench can be opened, a few plants can be removed, and the trench can be closed again.

Garden Storage

Here's the method the lazy man prefers.

It works well with certain crops, simply leaving them in the ground until they are needed. For some, the protection of a blanket of hay or straw or leaves is advised. For others, even this is not needed.

Let's run down through a list of the vegetables that can be stored in this way.

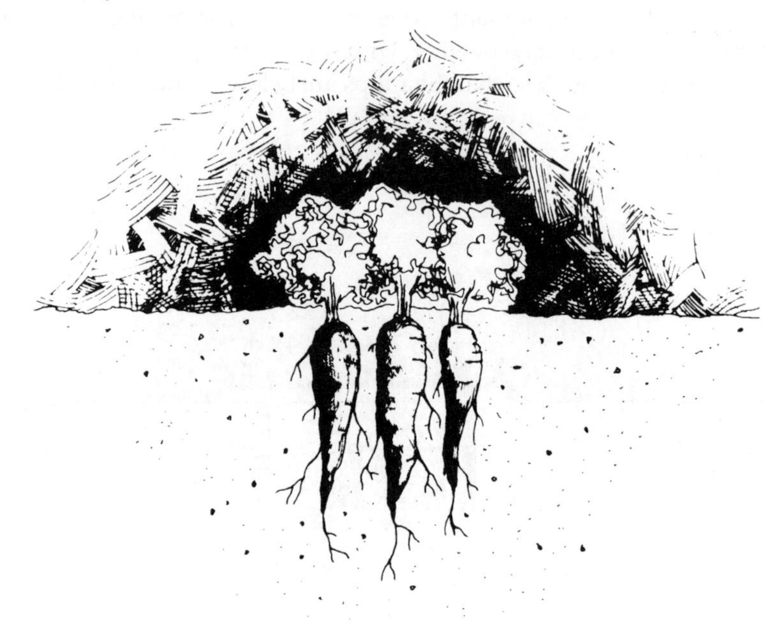

But, before we do that, one suggestion. Plan on this storage when you are planting your garden. Some of these crops should be planted in mid-summer so they mature in the late fall. Try to keep all of the "storage" crops in one area of your garden. It makes for easier work in the fall, when you should be cleaning out your garden, rototilling it, and planting a cover or green manure crop. If all of those storage crops are in one area, cleaning will be much easier. And so will harvesting, in the dead of winter when finding the vegetables can be a task.

Kale is a hardy vegetable and can be harvested and used like spinach well past Christmas in even the coldest parts of the country. Just let it grow.

Jerusalem artichokes are poor "keeper," but keep well if left undisturbed in the ground. Dig them when you want them. If the digging promises to be hard in colder areas, put down a layer of hay or leaves, and the digging will be easier. You'll miss a few of the 'chokes as you harvest them, and those will provide you your next season's crop.

Cabbage can be left in the garden until well after the first frosts. Halloween or severe freezes may finish the cabbages. Brussels sprouts will produce long after the first frosts, too.

For winter harvesting of the root crops, spread a heavy layer of leaves, hay, or straw over them. This prevents the possibility of alternate freezing and thawing, which can injure them, and makes it easy to retrieve them when they're wanted.

Mark both ends of each row with stakes. It will simplify finding the crops after a snowstorm.

Notice the flavor of these crops as you harvest and eat them. We think some improve with the cold. These include carrots, parsnips, salsify, and turnips. See whether you agree.

Storage Crops

Apples

Store apart from vegetables, since apples may pick up some flavors, as from potatoes and onions, and may make carrots bitter. Keep temperatures as close as possible to 32°F. Lawrence Southwick, a veteran grower, recommends cooling picked apples as quickly as possible. Root cellar conditions are ideal for apples. Don't try to store windfalls or bruised fruit. They will only spoil the others. Apples that mature in late fall store best. (See table for keeping quality of varieties.) Apples are stored best in creates that can be stacked on top of each other. Those who store applies in deep bins find the weight of the pile will damage those on the bottom, causing them to decay.

Beets

Harvest after first light frosts, let dry in sun, then cut off tops, leaving at least two inches of tops so beets won't bleed. Discard any that are immature, damaged, or show signs of decay. Store as described for carrots. Beets do not store as well as carrots, so plan to use them as quickly as possible.

Cabbages

For keeping, grow winter cabbages. Cabbages should be mature, and feel heavy for their size. Their odor can permeate a home, so if you place them in your root cellar, cut off the root and outer leaves, then wrap them in several layers of newspaper. Other ways to store: Outside in a pit or trench, or hung up by the root in shed or garage, or simply left in the garden, to harvest before severe frosts.

Carrots

Most varieties store well, the thicker varieties keep best. Can be heavily mulched in the garden, and dug up all winter. Or dig up, let dry in the sun, then store in boxes in root cellar. One method is to put a heavy layer of sand in the box, then layer carrots (they can touch, but don't jam too many together), more sand, and another layer of carrots, repeating until box is filled. Sand must be kept damp to keep carrots in good, unshriveled condition. They'll also keep well in outside pits.

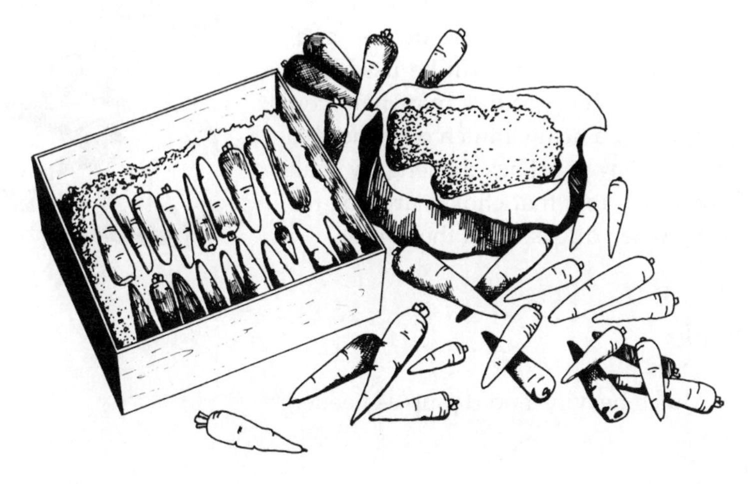

Celeriac

This lesser-known cousin of celery is nutritious, tasty, and a good keeper. Start a late crop to harvest in the late fall, then store as described for carrots.

Celery

Giant Pascal is a good keeper. Use trench method described for outside storage. Or take up plants with roots, pack them in a box with roots in soil, and store in root cellar. Keep sand or soil moist, but keep water off the tops. In areas having temperate winters, leave plants in the ground, mulch heavily with leaves or hay, and harvest as needed.

Garlic

Harvest and keep like onions. Garlic can add beauty to your home if braided and hung on the wall.

Jerusalem Artichokes

Dug up, they're poor keepers, no matter what method you try. But they store just fine if left in the ground. Dig them up when they're needed — and dig up no more than will be used in a week. If frozen ground may halt your digging, try a heavy mulch over the area, after you have cut back the lengthy stalks. Remember, too, that chokes are fine for eating when dug early in the spring.

Leeks

Mulch heavily, and dig up as needed.

Onions

The stronger the taste, the better they keep. Late-maturing onions will keep best. Also those with thin necks. Good keepers include Ebenezer, Southport, and Yellow Globe. Harvest when most of the vegetation has fallen over. Let them cure for several weeks in a warm, dry, and well-ventilated area. Can be in the sun. Skins will rustle when they're ready to be stored. Use, don't store, those with wide necks or green stalks. Place in open-weave bags or open slatted crates, in cool, dry room. Attics are often ideal.

Parsnips

Can be stored like carrots in root cellar or, better still, a pit. Far easier to leave them in the ground, mulch heavily, and dig them when they're needed. Cold weather improves their taste, so they're best when dug in the early spring.

Popcorn

Harvest when the stalks and leaves are dry. Let ears cure for about three weeks in a warm, dry area, then shell by twisting off kernels with hands. If kernels cling, let the ears cure longer. Store kernels in closed jars in room temperature. For decorations — as well as eating, pull back the husks, braid them, and hang them in your home. Remove as needed for popping,

Potatoes

Harvest after most of the tops have died down, and potatoes are mature. Harvesting can be delayed as much as six weeks without harming the crop. Let potatoes dry on the ground for several hours after digging, then store them in the dark, to avoid having them turn green. Don't store them near apples. Pick out all immature potatoes, any that were damaged while digging, and any that show signs of rot.

Potatoes can be put in boxes. They need ventilation. Ideal temperatures are between 40° and 50°F. If stored at cooler than 40°F, potatoes tend to become sweet, but can be returned to original flavor by leaving them in room temperature for a week or two before eating them.

A method of outdoor storage is the potato mound. Around the area to be used, dig a shallow drainage ditch. Also dig an x of trenches crossing the bottom of the mound area. Cover the trenches with wire screening to keep out rodents, but allow air to enter and ventilate the mound. Construct a circular vent of screening that will extend vertically from the trenches to the top of the mound like a chimney.

The floor of the mound should be layered with one foot of clean hay. Pile the potatoes into the area around the vertical vent and over the cross-ventilation trenches. Cover the potatoes with about six inches of hay for insulation. A final insulating layer of soil about 6"–12" thick should be placed over the hay. The circular vertical vent should extend about six inches above the surface of the mound, and must be covered with a lid to keep out water. An inverted tin can works well. When the temperature falls below freezing, the cross-ventilation trenches should be blocked with soil. For a rough rule-of-thumb, pile the potatoes so that the height of the mound is about one-half its diameter.

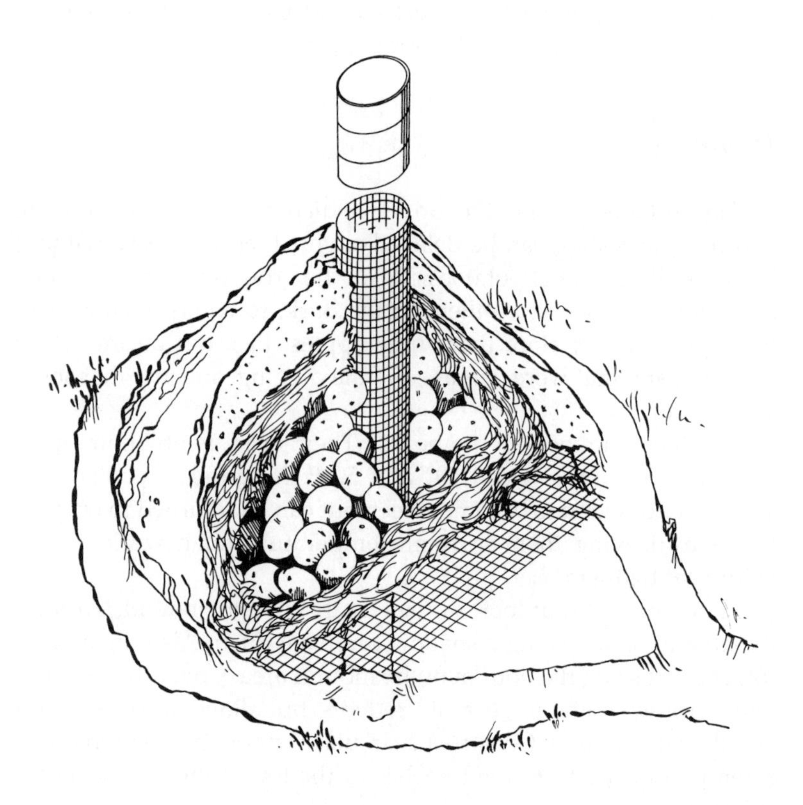

Pumpkins

Harvest and store as described under squash.

Rutabaga

Harvest before heavy frost. Must be kept moist to avoid shriveling. Follow directions for carrots.

Salsify

Can be stored as carrots, in root cellar or in outdoor pit. It's far easier to leave them in the ground, mulch heavily enough to keep ground from freezing, and dig up as wanted.

Squash

Good keepers are Hubbard, Butternut, Acorn. Leave on plant until fully matured, but pick before heavy frost, leaving a stem on the squash. Leave in sunny field for several weeks, or cure in a dry warm place for two weeks. They're ready to store when a fingernail won't penetrate the skin. Don't cure the Acorns. They may lose moisture, turn orange, or become stringy. Store squash on shelves, preferably not piled up, in warm (50°–60°F), dry area.

Sweet Potatoes

Harvest before frost. Cure potatoes in high humidity and 80°–85°F temperatures for two weeks to toughen skins. Then store in warm (55°–60°F), relatively dry area. Light won't affect sweet potatoes as it will their Irish brethren.

Tomatoes

Green tomatoes are well worth harvesting. Those in the light green stage will ripen, providing good eating long after the vines have been killed by frosts. Harvest before those frosts. Try one of several methods to save them. One is to pull up the entire plant, roots and all, and hang it upside down in a place where frost will not reach it. Or pick the light green tomatoes, wrap them in newspapers individually, then keep them in temperatures of 55°–60°F, where they will ripen during the next six weeks.

Table 1. Storage life expectancies, recommended storage temperatures, and relative humidities of fresh fruits *

Fruit	Temperature °F	(°C)	Relative humidity %	Approximate length of storage period
Apples	30	(-1.1)	85–90	as per variety
Apricots	32	(0.0)	85–90	1–2 weeks
Blackberries	same as raspberries			
Cherries				
Sweet	32	(0.0)	85–90	2–3 weeks
Sour	32	(0.0)	85–90	few days
Cranberries	36–40	(2.2–4.4)	80–85	2 months
Grapes, American	32	(0.0)	85–90	1 month
Peaches	32	(0.0)	85–90	2 weeks
Pears				
Bartlett	30	(-1.1)	85–90	2–3 months
Fall and winter	30	(-1.1)	85–90	3–5 months
Plums				
Early, Japanese type	40	(4.4)	85–90	few days
Other types	32	(0.0)	85–90	4–6 weeks
Raspberries	32	(0.0)	85–90	few days
Strawberries	32	(0.0)	85–90	5–10 days

* Based on information from Canada Department of Agriculture Publications 1532, *Storage of fruits and vegetables*, by S.W. Porritt.

Table 2. Storage life expectancies, recommended storage temperatures, and relative humidities of fresh vegetables *

Vegetable	Temperature °F	(°C)	Relative humidity %	Approximate length of storage period	Suggested methods for extended preservation
Asparagus	32	(0.0)	95	3 weeks	freeze or can
Beans					
green or snap	45–50	(7–10)	85–90	8–10 days	freeze or can
lima					
shelled	32	(0.0)	85–90	2 weeks	freeze or can
unshelled	32	(0.0)	85–90	2 weeks	
Beets					
bunched	32	(0.0)	90–95	10–14 days	
topped	32)	(0.0)	90–95	1–3 months	
Broccoli					
Italian or					
sprouting	32	(0.0)	90–95	1 week	freeze
Brussels sprouts	32	(0.0)	90–95	3–4 weeks	freeze
Cabbage					
early	32	(0.0)	90–95	3–4 weeks	
late	32	(0.0)	90–95	3–4 months	
Carrots					
bunched	32–34	(0.0–1.1)	95	2 weeks	
topped	32–34	(0.0–1.1)	95	4–5 months	
Cauliflower	32		90–95	2 weeks	freeze
Celery	32	(0.0)	95+	3 months	
Corn, sweet	32	(0.0)	90–95	8 days	freeze or can
Cucumbers	45–50	(7.2–10)	95	10–14 days	
Eggplants	45–50	(7.2–10)	85–90	10 days	
Endive or escarole	32	(0.0)	90–95	2–3 weeks	
Garlic, dry	32	(0.0)	70–75	6–8 months	
Horseradish	30–32	(-1.1–0.0)	90–95	10–12 months	can
Kohlrabi	32	(0.0)	90–95	2–4 weeks	freeze
Leeks, green	32	(0.0)	90–95	1–3 months	
Lettuce (head lettuce)	32	(0.0)	95	2–3 weeks	

continued					
Vegetable	Temperature °F	(°C)	Relative humidity %	Approximate length of storage period	Suggested methods for extended preservation
Melons					
Cantaloupe or					
muskmelon	32–45	(0.0–7.2)	85–90	2 weeks	
honeydew	45–50	(7.2–10)	85–90	2–3 weeks	
watermelons	36–40	(2.2–4.4)	85–90	2–3 weeks	
Mushrooms,					
cultivated	32	(0.0)	85–90	5 days	freeze
Onion sets	32	(0.0)	70–75	5–7 months	
Onions, dry	32	(0.0)	50–70	5–9 months	
Parsnips	32	(0.0)	95	2–4 months	
Peas, green	32	(0.0)	95	1–2 weeks	freeze or can
Peppers, sweet	45–50	(7.2–10)	85–90	8–10 days	freeze
Potatoes					
early-crop	50	(10)	85–90	1–3 weeks	
late-crop	39	(3.9)	85–90	4–9 months	
Pumpkins	45–50	(7.2–10)	70–75	2–3 months	
Radish					
spring, bunched	32	(0.0)	90–95	2 weeks	
winter	32	(0.0)	90–95	2–4 months	
Rhubarb	32	(0.0)	90–95	2–3 weeks	freeze
Rutabaga or turnip	32	(0.0)	90–95	6 months	
Salsify	32	(0.0)	90–95	2–4 months	
Spinach	32	(0.0)	90–95	10–14 days	freeze or can
Squash					
summer	45–50	(7.2–10)	70–75	2 weeks	
winter	45–50	(7.2–10)	70–75	6 months	
Tomatoes					
ripe	50	(10)	85–90	3–5 days	
mature green	55–60	(12.8–15.6)	85–90	2–6 weeks	

* Based on information from Canada Department of Agriculture Publications 1532, *Storage of fruits and vegetables*, by S.W. Porritt.

Table 3. Normal and maximum storage periods for some common apple varieties *

Variety	Storage Period	
	Normal months	Maximum months
Wealthy	0–1	3
Grimes Golden	2–3	4
Jonathan	2–3	4
McIntosh	2–4	4–5
Cortland	3–4	5
Spartan	4	5
Rhode Island Greening	3–4	6
Delicious	3–4	6
Stayman	4–5	5
York Imperial	4–5	5–6
Northern Spy	4–5	6
Rome Beauty	4–5	6–7
Newton	5–6	8
Winesap	5–7	8

* Based on information from Canada Department of Agriculture Publications 1532, *Storage of fruits and vegetables,* by S.W. Porritt.

Building an Underground Root Cellar

Building an underground root cellar is a fairly ambitious project. But if you don't have the right basement for a cold storage room, if pits and trenches don't provide enough storage space, or if you want a room that can double as a storm shelter, then you probably want to build an underground root cellar.

Storey Publishing Bulletin A-76, *Build Your Own Underground Root Cellar* gives detailed, step-by-step instructions for building an 8' x 12' concrete block root cellar.

The U.S.D.A. also has plans for underground root cellars. See your county extension agent for more details.

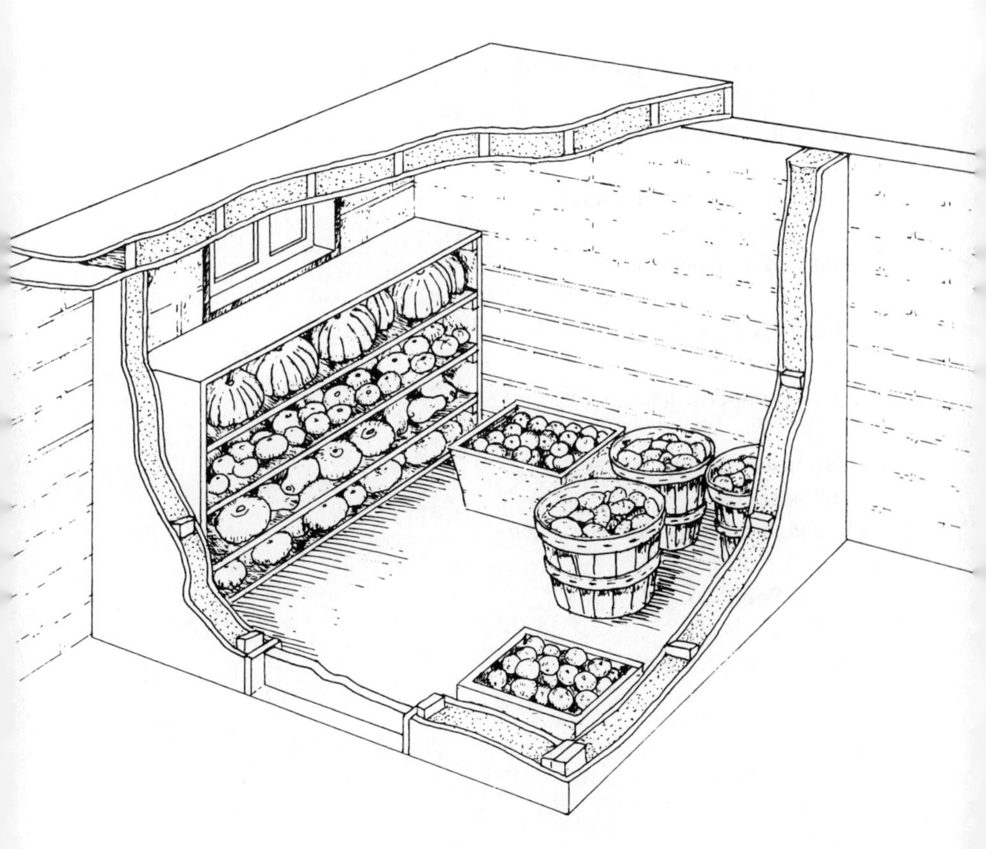

Seed Sources

Many long keeping varieties of garden vegetables are available through the Seed Savers Exchange. Visit the Seed Savers website at *www.seedsavers.org* or call them at 563-382-5990.

Vegetable Seeds

Alberta Nurseries
403-224-3544
www.gardenersweb.ca

Gurney's Seed & Nursery Co.
513-354-1492
www.gurneys.com

High Mowing Organic Seeds
802-472-6174
www.highmowingseeds.com

Johnny's Selected Seeds
877-564-6997
www.johnnyseeds.com

Jung Quality Garden Seeds
800-297-3123
www.jungseed.com

Nichols Garden Nursery
800-422-3985
www.nicholsgardennursery.com

Park Seed Co.
800-845-3369
http://parkseed.com

Pinetree Garden Seeds
207-926-3400
www.superseeds.com

Seeds of Change
888-762-7333
www.seedsofchange.com

Stokes Seeds
800-396-9238
www.stokeseeds.com

Territorial Seed Company
800-626-0866
www.territorialseed.com

Vesey's Seeds Ltd.
800-363-7333
www.veseys.com

W. Atlee Burpee & Co.
800-888-1447
www.burpee.com

Seed Potatoes

Some general catalogs also carry seed potatoes.

Eagle Creek Seed Potatoes
877-224-3939
www.seedpotatoes.ca

Ronniger Potato Farm
877-313-7783
www.potatogarden.com

Wood Prairie Farm
800-829-9765
www.woodprairie.com

Other Storey Titles You Will Enjoy

Put'em Up! by Sherri Brooks Vinton
A comprehensive guide to preserving: bright flavors,
flexible batch sizes, modern methods.
304 pages. Paper. ISBN 978-1-60342-546-9.

Recipes from the Root Cellar by Andrea Chesman
A collection of more than 250 recipes for winter kitchen produce —
jewel-toned root vegetables, hardy greens, sweet winter squashes,
and potatoes of every kind.
400 pages. Paper. ISBN 978-1-60342-545-2.

Root Cellaring by Mike and Nancy Bubel
Suitable for city and country folks, with information on harvesting and
creating cold storage anywhere — even closets! — plus 50 recipes.
320 pages. Paper. ISBN 978-0-88266-703-4.

Saving Vegetable Seeds by Fern Marshall Bradley
This illustrated, step-by-step guide shows you how to save seeds
from 20 of the most popular vegetable garden plants, including
beans, carrots, peas, peppers, and tomatoes. You'll learn how each
plant is pollinated, how to select the seeds to collect, and how to
process and store collected seeds.
80 pages. Paper. ISBN 978-1-61212-363-9.

Starting Seeds by Barbara W. Ellis
A simple and straightforward guide that will remove
the stress and mystery of seed starting.
128 pages. Paper. ISBN 978-1-61212-105-5.

***The Vegetable Gardener's Bible,* 2nd edition**
by Edward C. Smith
The 10th Anniversary Edition of the vegetable gardening classic,
with expanded coverage of additional vegetables, fruits, and herbs.
352 pages. Paper. ISBN 978-1-60342-475-2.
Hardcover. ISBN 978-1-60342-476-9.

These and other books from Storey Publishing are available
wherever quality books are sold or by calling 1-800-441-5700.
Visit us at *www.storey.com* or sign up for our newsletter
at *www.storey.com/signup.*